Flávia Gonçalves Fernandes

Electronics

Flávia Gonçalves Fernandes

Electronics

Reports for Practical Classes

ScienciaScripts

Imprint

Any brand names and product names mentioned in this book are subject to trademark, brand or patent protection and are trademarks or registered trademarks of their respective holders. The use of brand names, product names, common names, trade names, product descriptions etc. even without a particular marking in this work is in no way to be construed to mean that such names may be regarded as unrestricted in respect of trademark and brand protection legislation and could thus be used by anyone.

Cover image: www.ingimage.com

This book is a translation from the original published under ISBN 978-613-9-66132-9.

Publisher:
Sciencia Scripts
is a trademark of
Dodo Books Indian Ocean Ltd. and OmniScriptum S.R.L publishing group

120 High Road, East Finchley, London, N2 9ED, United Kingdom
Str. Armeneasca 28/1, office 1, Chisinau MD-2012, Republic of Moldova, Europe
Printed at: see last page
ISBN: 978-620-8-01374-5

SUMMARY

SUMMARY

Analogue electronics is the branch of electronics that studies the performance of analogue devices and circuits such as resistors, capacitors, coils, potentiometers, transistors, crystals and integrated circuits for the most part. It also studies the behaviour of electrical signals such as radio frequency (AM and FM signals), Bluetooth, Wireless, etc. It can also be defined as an area of electronics that studies physical characteristics in an analogue way to pre-established electrical references such as: temperature, weight measurement, voltage and current measurement (Multimeter), among other things. According to an unknown author, "Analogue electronics developed with the advent of the control of physical quantities, variable or not, oscillatory forms, at low or high frequencies, and which are used in almost all types of equipment[...]", in other words, the study of analogue electronics arose with the need to control certain quantities so that they could be used or converted into real values or to use and convert real values into quantities.

According to the English Wikipedia page: "Analogues are electronic systems with a continuously varying signal, in contrast to digital electronics, where the signals assume only two levels. The term analogue describes the proportional relationship between a signal and a voltage or current that represents the signal."

The first traces of electronics occurred in 1837 with the creation of the telegraph by Samuel Morse in the USA. In 1879, Thomaz Edison experimented to create the first incandescent light bulb in history. Later, John Ambrose Flemming created a device that consisted of wrapping the filament of an electric bulb around a cylindrical plate, which he called an electric valve (or diode, as it later became known). At the end of the 1930s, more modern valves began to appear, such as: directed beam valves, cathode ray tubes, tuning tubes, metal valves and miniature valves. As a consequence, doped crystal diodes also appeared, which later gave rise to the transistor in the 1950s.

Applications

Analogue electronics is the basis for areas such as telecommunications, power electronics, digital electronics and microelectronics, among others. The main circuits created and studied with analogue electronics are: controlled and uncontrolled rectifiers (half-wave, full-wave and bridge), polarising circuits, voltage limiters and regulators, amplifier circuits, timers and oscillators, sensors and couplers.

Rectifiers

Rectifiers are circuits that transform alternating current into direct current. They are also known as AC/DC converters. There are two types of rectifiers: non-controlled and controlled. Uncontrolled rectifiers use ordinary diodes to convert alternating current into direct current. Controlled rectifiers use thyristors (SCR, DIAC, TRIAC, PUT and SCS, being the main ones) that control the rectification firing angle so that, as well as converting AC into DC, there is power control in inductive loads such as motors.

Polarisers, limiters and voltage regulators

Polarisers and voltage regulators are circuits that use transistors to determine voltage and/or current values that remain stable so that they can operate at a certain working temperature. Voltage limiters are circuits that use diodes (most commonly Zener diodes) to protect the load from excess voltage.

Operational amplifier as differential amplifier.

Amplifiers

The amplifier is a circuit that uses transistors (BJT, FET, JFET and MOSFET being the most common) to amplify an analogue signal (voltage, current, audio, AM signal, etc). In other words, a power converter. The input signal only controls the current flowing from the power supply or battery. Thus, the energy from the power supply is converted by the amplifier into a power signal.

Operational Amplifiers

Operational amplifiers, or amp-ops, are amplifiers with very high input impedance (ideally infinite) and very low output impedance (ideally zero) and very high gain. The most common amp-op uses are: non-inverting amplifier, unit follower, adder, integrator and differentiator.

Timers and oscillators

Timers and oscillators use the CI LM 555 to control the output (on or off). Timers are circuits that send an output signal for a certain period of time. You can control a load automatically for a certain time. Oscillators are circuits that create an oscillatory or sinusoidal signal.

Sensors

Sensors are components whose output signal varies depending on its exposure. Examples: presence sensors, temperature sensors (LM35), humidity sensors (HIH-4000-001), among others.

CHAPTER 1

THE TRANSISTOR AS A KEY

1 . OBJECTIVES

Present the basic operation of the bipolar junction transistor, schematic symbol, operating region and limits, data sheet and its application as a "switch".

2 .THEORETICAL CONCEPTUALISATION

2.1. The transistor

The transistor plays the role of amplifier. When directly polarised, the diode conducts electricity. The transistor introduces a new capability, which is the possibility of controlling how much electricity is conducted.

It all starts when you add an additional layer to a diode. Instead of two portions, P and N, of silicon, let's see what happens when we add three portions, making a sandwich of an N portion, as can be seen in Figure 1 below.

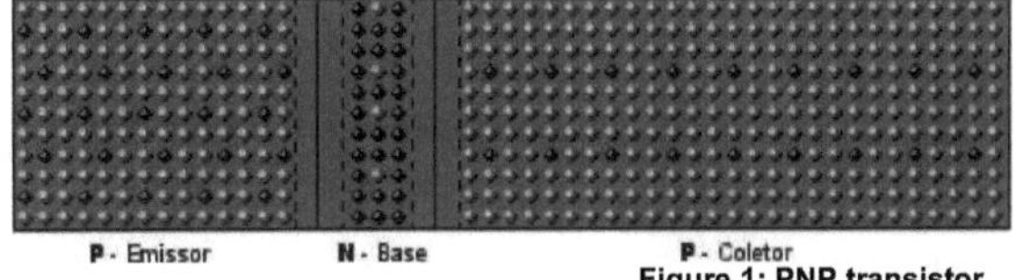

Figure 1: PNP transistor.

Each of these layers has its own peculiarities:

- the first P layer (on the left) has a medium width and is heavily doped, i.e. it has many trivalent atoms. This makes this layer a supplier of gaps (positive charges). It is therefore called an emitter.

- the N centre layer is very thin and has a medium doping. As it is thin, it does not represent a major obstacle to the charges coming from the emitter. This layer is called the base.

- the layer on the right is quite large compared to the others and is weakly doped. Because it is responsible for receiving the electrons that leave the emitter and pass through the base, this layer is called the collector.

As you can see in the figure, the transistor is the result of joining two diodes. Except that both diodes share the base, i.e. the N element in this case. There are therefore two junctions. As you might expect, a depletion layer is formed at each of the junctions, where the electrons and gaps are balanced, creating a potential barrier.

2.2 Transistor operation

Firstly, we'll place a battery between the emitter and the base. To polarise it directly, we connect the negative terminal (electron flow) of the battery to the emitter (N portion - excess electrons) and the positive terminal (gap flow) to the base (P portion - excess gaps). In this way, the N region, with excess electrons, receives even more electrons, and the P portion receives even more gaps, as seen in Figure 2.

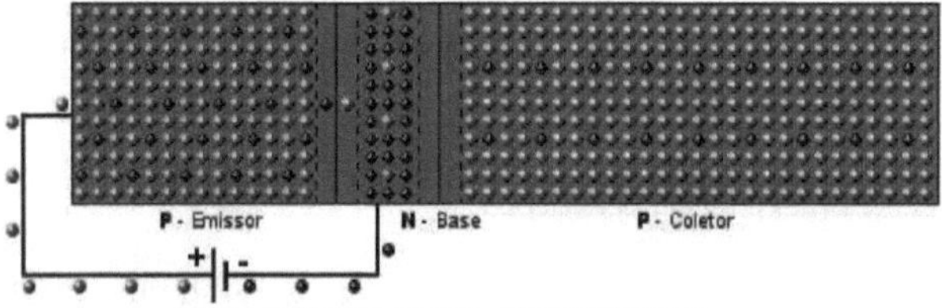

Figure 2: PNP transistor - electron transfer.

Direct polarisation makes the base-emitter portion behave exactly like a conductor.

At the same time, we are going to reverse polarise the base-collector assembly. To do this, we connect the positive terminal (flow of gaps) of the battery to the collector (N portion - excess electrons) and the negative terminal (flow of electrons) to the base (P portion - excess gaps). In this way, the electrons in the collector will be attracted to the gaps in the battery's positive pole and the gaps in the base will be filled by the electrons in the negative pole. This reverse polarisation means that the base-collector portion does not conduct current.

•   If we do the two previous polarisations simultaneously, we get Figure 3, where in the base-emitter polarisation (the first one that we saw), the electrons were travelling towards the base, attracted by the positive pole of the battery. But now the collector, which is much larger and has extra energy coming from the negative pole of the battery, exerts a much greater attraction on these electrons. As the base is very thin, the electrons tend to pass through the base and go to the collector rather than flow through the base to the positive pole of the battery. In this way, a small part of the current will flow through the base; most of the current will flow to the collector.

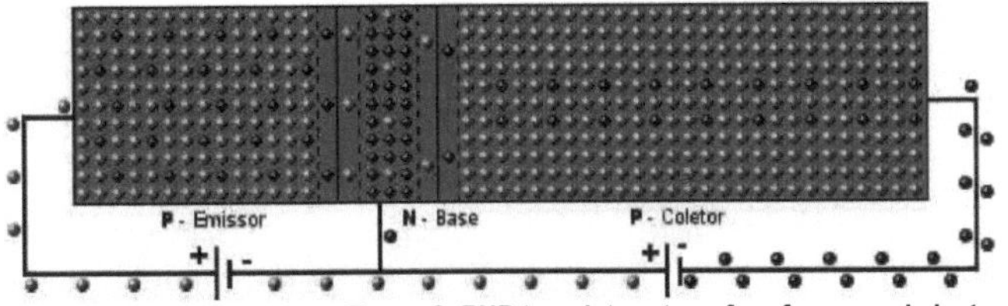

Figure 3: PNP transistor - transfer of gaps and electrons.

This is because:

-if   we increase the current flowing through the base (emitter-base), there will be an increase in the current flowing through the collector. In other words, we can control the current coming from the emitter to the collector by acting on the base current. In other words: the base current controls the current between the emitter and collector.

-since   the base current is very small, we only need to apply a small variation in the base current to obtain a large variation in the collector current. That's it: we enter with a small current (via the base) and leave with a large current (via the collector).

We have analysed a PNP transistor. If we invert our sandwich, we create an NPN transistor. The operation is exactly the same, only the current flow is reversed.

In electronic schematics, which are "maps" of how an electronic circuit is designed, transistors are represented by the following symbols in Figure 4:

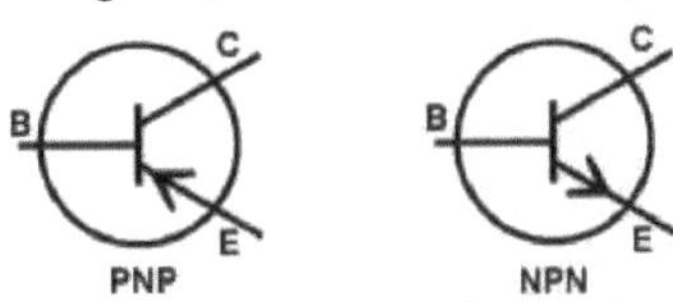

Figure 4: Transistor schematic symbols.

There are a number of devices that work based on precisely this principle, such as the microphone and the loudspeaker, visualised below in Figure 5, for example. It's electrons flowing, enabling everything we know as *electronic.* That was the beginning of our Digital Age.

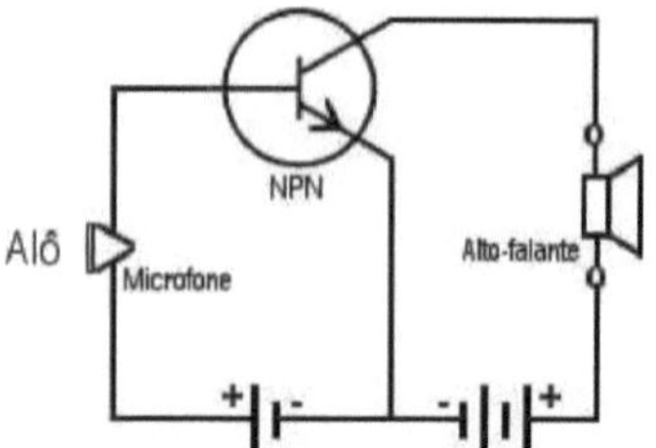

Figure 5: Schematic of transistor amplification.

3 . EQUIPMENT AND PERMANENT MATERIALS REQUIRED

The following equipment and permanent materials were needed to carry out this experiment:

- Function generator;
- Oscilloscope - 2 channels;
- BC547 transistor;
- Protoboard;
- DC supply of +12V;
- 2.7 kΩ and 820 kΩ resistors;
- 1 led;
- Cables for connections.

4 .EXPERIMENTAL PROCEDURE

Firstly, using the data sheet for the BC547 transistor, its encapsulation is shown in Figure 6 below, with its respective terminals: base, collector and emitter.

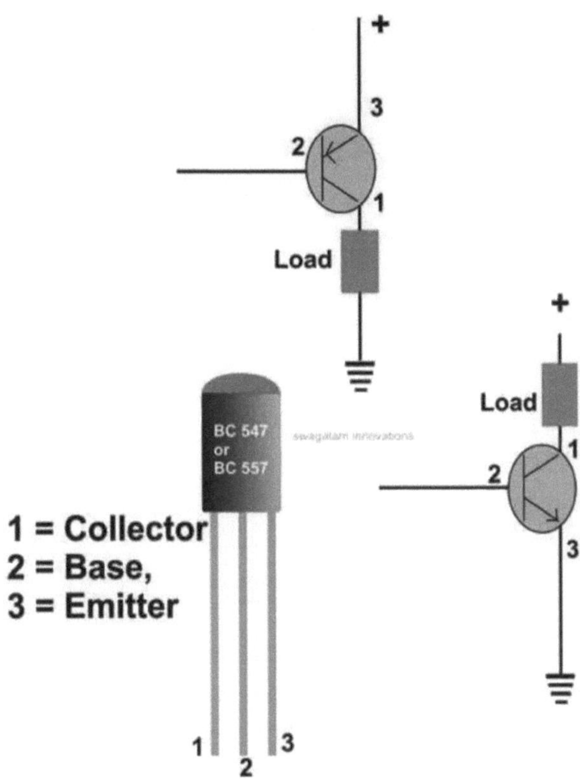

Figure 6: Representation of the BC547 transistor package.

Subsequently, using the multimeter, the Bcc of the BC547 transistor received was measured and a voltage of 293 V was found. The input signal Vi was then adjusted using the function generator: square wave with amplitude V = 5 Vp and frequency 1kHz. The input signal found is shown in Figure 7 below.

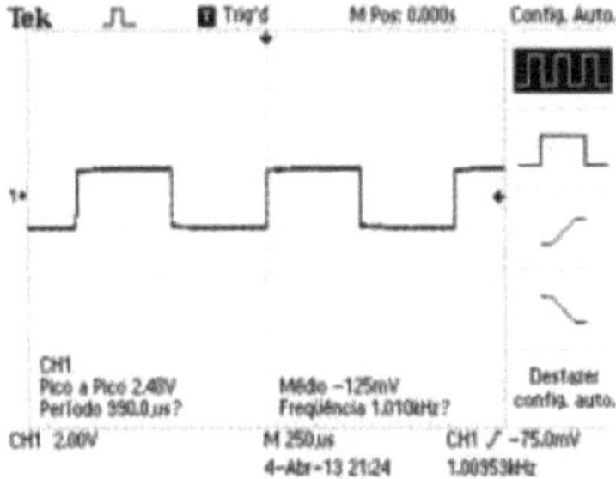

Figura 7:   Oscilloscope input signal (1A) - V/div = 2V/div - Time/s = 250ps.

The voltage of the Vcc source was then adjusted to 12 VDC and measured with a multimeter to confirm that it was approximately 12 V. The circuit shown in Figure 8 was then assembled on the protoboard.

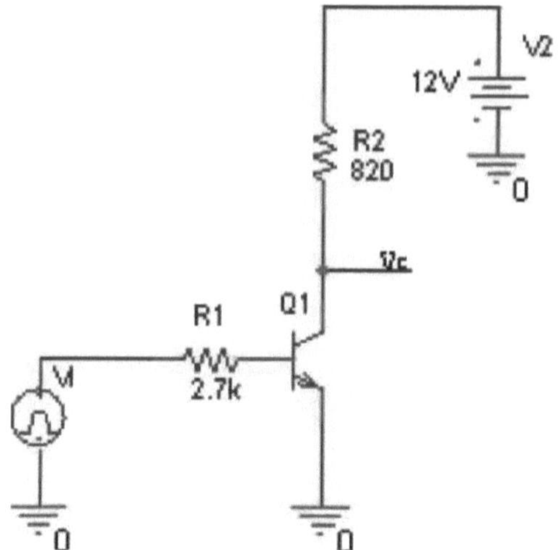

Figure 8: Circuit assembled on the protoboard during the experimental procedure.

After assembling the circuit, the output signal was found and represented in Figure 9 below, where the value Vp = 12 V was found, which expresses the output signal of the source voltage Vcc, which also has a value of 12 V.

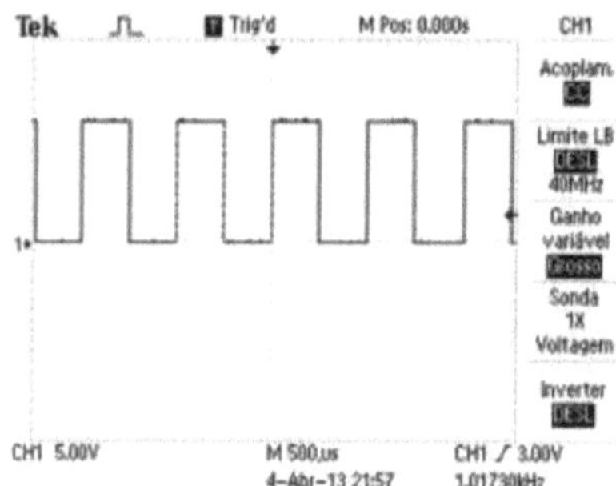

Figura 9: Output signal on oscilloscope (1B) - V/div = 5V/div - Time/s = 500ps.

Then channel 1 of the oscilloscope was connected to the input and channel 2 was connected to the output of the circuit. The signals found were plotted simultaneously in Figure 10 below. This is because when a pulsed signal (square wave) is applied to the base of the transistor, the transistor will operate like an electronic switch. When the voltage applied to the base of the transistor is low (0V), the transistor will not conduct any current, so there will be no current in R and the output voltage will be equal to the battery voltage (Vcc). When the voltage applied to the base of the transistor is high (5V), the transistor conducts and the output voltage is equal to the reference voltage (ground), i.e. 0V. So when a high-level voltage is applied to the control (signal) terminal, the switch closes, and when a low-level voltage is applied to the control (signal) terminal, the switch opens.

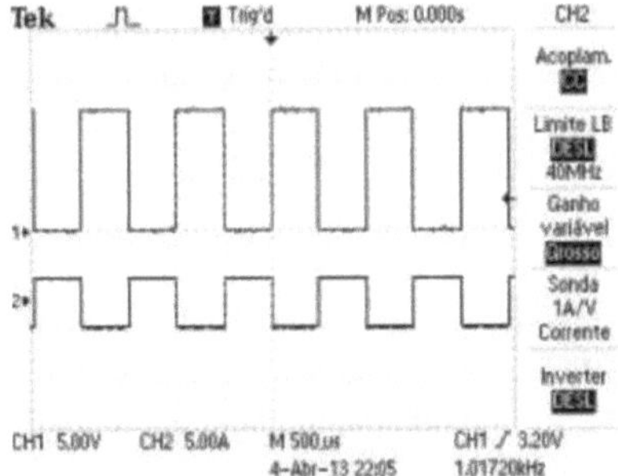

Figura 10: Input and output signals on the oscilloscope (1C) - V/div = 5V/div - Time/s = 500µs.

Finally, a light-emitting diode (LED) was placed in series with resistor R2 in the circuit and the frequency was lowered to 3 Hz. As a result, the LED flashes according to the "switched" signal, i.e. when the transistor conducts electric current, the LED lights up and when it doesn't conduct current, it goes out, in a very fast process. This is why the frequency has been reduced, because if the frequency is higher, this change in the LED's state is practically imperceptible.

5. RESULTS AND CONCLUSIONS

Therefore, the objective of the experimental lesson was achieved, as we acquired theoretical and practical knowledge about the basic functioning of the bipolar transistor, as well as learning how to use it and operate it correctly when assembling the circuit, which is fundamental for continuing with the syllabus of the Electronics subject.

Therefore, the assemblies and measurements requested in the script for this practical lesson were carried out successfully, and the results obtained from the experiment were as expected, given that we know how the transistor works as a "key" in the electrical circuit.

Finally, the practical procedure was carried out as a team, since there are several tasks and stages to be followed. By dividing up the roles, the work is more effective and the environment is more harmonious and fruitful for developing the proposed activities and ensuring that they are carried out on time, especially when the members of the group have a good relationship with each other, and learning becomes more effective and dynamic, since all the members share knowledge and experiences relating to the subject.

6. BIBLIOGRAPHY

BOYLESTAD, Robert Louis; NASHELSKY, Louis. **Electronic Devices and Circuit Theory.** 6 ed. vol. II. Rio de Janeiro: Prentice Hall, 1996.

_ . **Datasheet of the BC547.** Available at: < http://www.play.com.br/datasheet/BC547.pdf >. Accessed on: 04 April 2013.

_ . **How the transistor works.** Available at:

<http://www.agostinhorosa.com.br/artigos/funcionamento-do- transistor.html>.

Accessed on: 04 April 2013.

MALVINO, Albert Paul. **Electronics.** 4 ed. São Paulo: Makron Books, 1989.

CHAPTER 2

REPORT 2: BJT - LOAD LINE

1. OBJECTIVES

- Practical assembly with a transistor;

- Study the importance of choosing the operating point;

- Visualise the amplification of a sinusoidal signal with and without distortion;

- Effect of the slope of the load line on the voltage gain.

2.THEORETICAL CONCEPTUALISATION

The load line

The load line is a line that cuts through the collector characteristic curves to show each of the operating points of a transistor.

Why do we use the load line? Because it contains all the possible operating points of this circuit. When the base resistance varies from zero to infinity, the collector current and collector-emitter voltage vary. If we plot each pair of Ic and V$_{CE}$ values, we get a sequence of operations that rest on the load line.

To summarise: The load line is a visual representation of the possible operating points of a transistor and can be seen in Figure 1.

Saturation Point: Point where the load line intersects the saturation region of the collector curves.

Cut-off point: Point where the load line intersects the cut-off region of the collector curves.

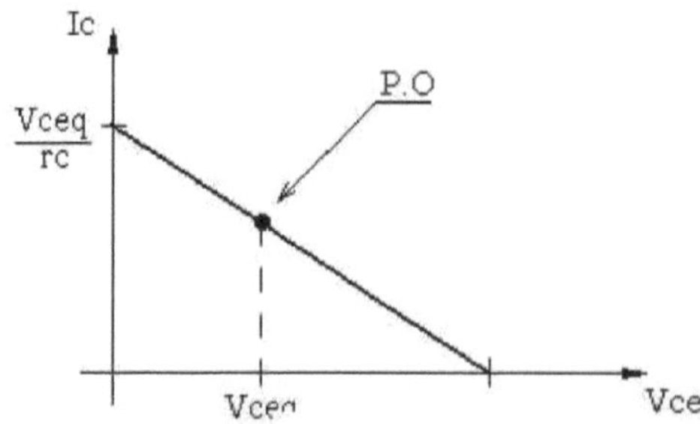

Figure 1: Bipolar Junction Transistor (BJT) Load Straight.

Operating Point (Quiescent): For transistor amplifiers, the resulting DC current and voltage establish an operating point on the characteristic curves that define the region that will be employed for amplifying the applied signal.

3. EQUIPMENT AND PERMANENT MATERIALS REQUIRED

The following equipment and permanent materials were needed to carry out this experiment:

- Oscilloscope OS-21 - 2 channels;
- Electrolytic capacitor - 2 units of 100uF (16V);
- Tips - 2 pairs (red and black);
- Voltage source;
- 1.2kΩ, 180kQ and 3.3kΩ resistors;
- Protoboard;
- BC547 transistor;
- Nose pliers, wire cutters;
- Multimeter
- Function generator;
- Cables for connections.

4.EXPERIMENTAL PROCEDURE

Firstly, the circuit in Figure 2 was assembled on the protoboard. The circuit was then fed with a voltage source of 12V and the function generator was set to a signal with a frequency of 500Hz and 5mVp.

The dc base current was then calculated using the base loop equation, remembering that the capacitor for direct current behaves like an open circuit, as follows:

$V1 - R_{BX} I_B - V_{BE} = 0$ considering $R_B = R2$

12-180kx I_B-0.7 = 0

I_B = 11.3/180k

I_B = 62,78uA

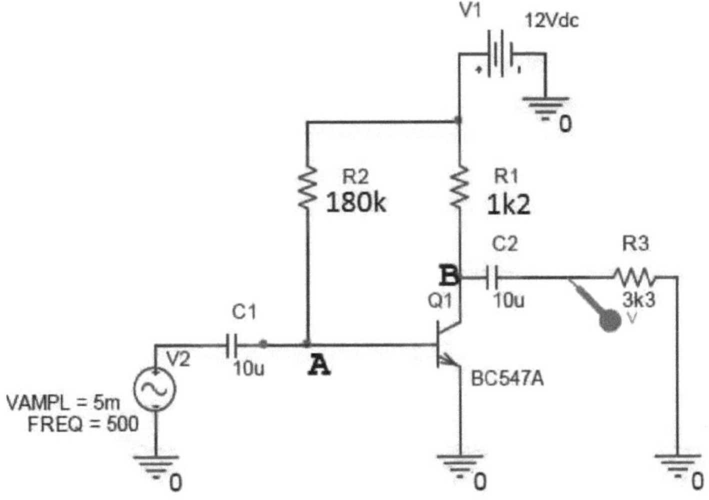

Figure 2: Circuit assembled on the protoboard.

In addition, the maximum collector current (ICsat) was calculated, considering VCE equal to zero, i.e. Ic = Vcc/Rc, as follows:

$V1 - Rc \times Ic - V_{CE} = 0$ considering $Rc = Ri$

12 - 1.2kx Ic-0 = 0

Ic= 12 / 1.2k

Ic = 10mA

The load line in Figure 3 below was also drawn, adopting the following points: (V_{CE} = 0, Icsat) and (V_{CE} = Vcc, Ic = 0).

15

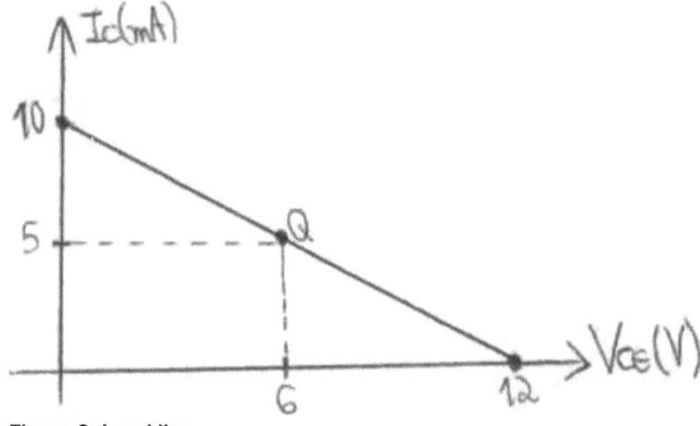

Figure 3: Load line.

Next, the function generator was set to a signal with a frequency equal to 1kHz and 5mVp. The output voltage on resistor R3 was then measured and the input and output signals are shown in Figures 4 and 5 respectively.

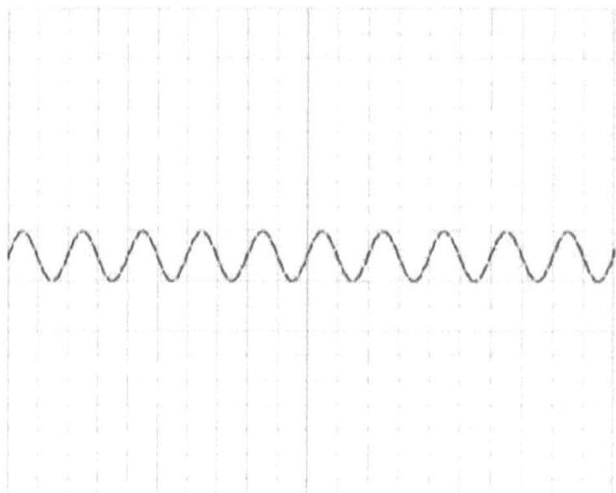

Figure 4: Oscilloscope input signal (1 A) - V/div = 5mV/div - Time/s = 500µs/div.

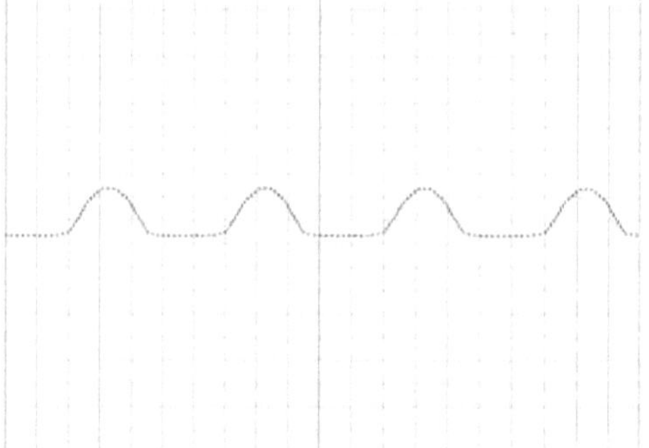

Figure 5: Oscilloscope output signal (1B) - V/div = 500mV/div - Time/s = 200µs/div.

Based on the voltage values found, the voltage gain, which represents the amplifier's performance, was calculated using the formula $G_V = V_{OUT}/V_{IN}$:

$$G_V = 100mV / 1 0mV \rightarrow G_V = 100$$

In addition, the value of V_{IN} was changed to 50mV and 500mV, noting that as this value was increased, it became closer until it became-.

This is practically equal to the horizontal line corresponding to the time axis. Thus, in the first case, the output signal increases its amplitude considerably, as well as suffering signal distortion. In the second case, the output signal is not only distorted, but also inverted, so that its Vp is turned downwards, i.e. negative. As a result, the voltage gain decreases. This is because the signal supplied to the amplifier's input produces an amplified output signal that increases with an increase in the input signal up to a certain limit. A signal with a high amplitude can produce saturation or clipping of the transistor, or both, and then the signal is clipped at the bottom or top, or both.

The V_{IN} frequency was then changed to 100kHz and it was observed that the output signal became jittery, but remained with the same initial settings. This is because the change in the amplifier's voltage gain is produced as a result of the circuit's unequal response to signals of different frequencies. This phenomenon is produced due to the presence of reactive components in the circuit, which have frequency-dependent behaviour (capacitances and inductances). In the medium frequency range, the gain is maximum. In the low frequency range, the circuit's coupling capacitances produce attenuation of the output signal, while in the high frequency range, attenuation is produced by the transistor's intrinsic capacitances and the circuit's parallel capacitances.

On the load line with no current through the transistor, the voltage drops across the BJT and, on short-circuit, the voltage is shared by the collector and emitter resistors.

The curves found are device-dependent, however, so you can't rely on the particular transistor to behave as described in the data sheet. That's why charge lines are really best used for developing physical understanding.

The collector output resistor connects the amplifier's output to its input, thus destroying the one-sided nature of the amplifier and complicating the analysis considerably. However, since the collector resistor is high, its inclusion in the analysis has little effect on the amplifier's performance and can be eliminated from the circuit.

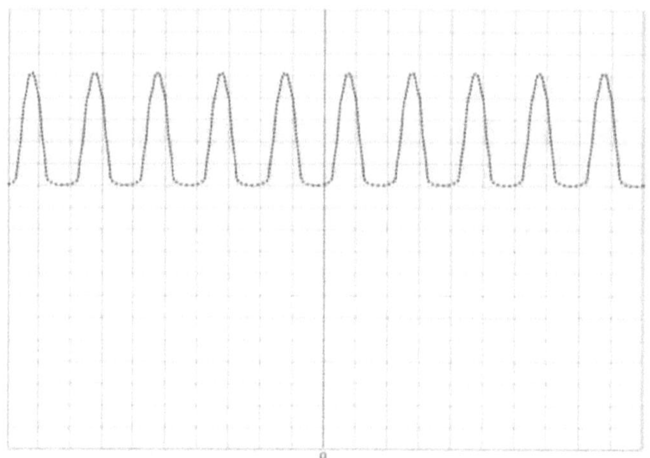

Figure 6: Output signal at the base of the transistor (1C) - V/div = 200mV/div - Time/s = 500µs/div.

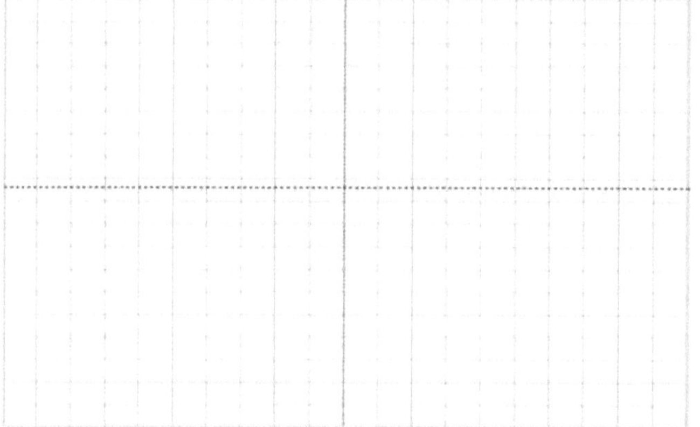

Figure 7: Output signal at the collector of the transistor (1 D) - V/div = 20V/div - Time/s = 500µs/div.

Thus, by increasing the resistance of the collector resistor, its current decreases and, consequently, the angle of inclination of the load line also decreases. And when you reduce its resistance, the opposite happens. If you change the resistance of the base resistor, there will be no interference with the current.

the load line, since it depends on I_C and V_{CE}, but the quiescent point will change due to the value of I_B found. However, by increasing the value of the base resistor, the signal is attenuated, i.e. the distortion of the amplifier's output signal is reduced.

The output signals at the base and collector were then checked on the oscilloscope and shown in Figures 6 and 7 respectively.

5. RESULTS AND CONCLUSIONS

Therefore, the objective of the experimental lesson was achieved, as we acquired theoretical and practical knowledge about the basic functioning of the bipolar transistor, as well as learning how to use it and operate it correctly when assembling the circuit, which is fundamental for continuing with the syllabus of the Electronics subject.

Therefore, the assemblies and measurements requested in the script for this practical lesson were carried out successfully, and the results obtained from the experiment were as expected, as we learnt how the transistor works as an amplifier in an electrical circuit, acting with and without distortion. We also learnt the importance of choosing the operating point and the effect of the slope of the load line on the voltage gain.

Finally, the practical procedure was carried out as a team, since there are several tasks and stages to be followed. By dividing up the roles, the work is more effective and the environment is more harmonious and fruitful for developing the proposed activities and ensuring that they are carried out on time, especially when the members of the group have a good relationship with each other, and learning becomes more effective and dynamic, since all the members share knowledge and experiences relating to the subject.

6. BIBLIOGRAPHY

BOGART Jr, **Theodore F, Electronic Devices and Circuits,** Volume I, 3º edition, São Paulo: MAKRON Books, 2001

BOYLESTAD, Robert Louis; NASHELSKY, Louis. **Electronic Devices and Circuit Theory.** 6 ed. vol. II. Rio de Janeiro: Prentice Hall, 1996.

_ . **Datasheet of the BC547.** Available at: < http://www.play.com.br/datasheet/BC547.pdf >. Accessed on: 04 April 2013.

MALVINO, Albert Paul. **Electronics.** 4 ed. São Paulo: Makron Books, 1989.

CHAPTER 3

BJT SIMULATION - LOAD LINE

1. OBJECTIVES

- Perform transistor simulation;
- Study the importance of choosing the operating point;
- Visualise the amplification of a sinusoidal signal with and without distortion;
- Effect of the slope of the load line on the voltage gain.

2 .THEORETICAL CONCEPTUALISATION

The load line

The load line is a line that cuts through the collector characteristic curves to show each of the operating points of a transistor.

Why do we use the load line? Because it contains all the possible operating points of this circuit. When the base resistance varies from zero to infinity, the collector current and collector-emitter voltage vary. If we plot each pair of Ic and V_{CE} values, we get a sequence of operations that rest on the load line.

To summarise: The load line is a visual representation of the possible operating points of a transistor and can be seen in Figure 1.

Saturation Point: Point where the load line intersects the saturation region of the collector curves.

Cut-off point: Point where the load line intersects the cut-off region of the collector curves.

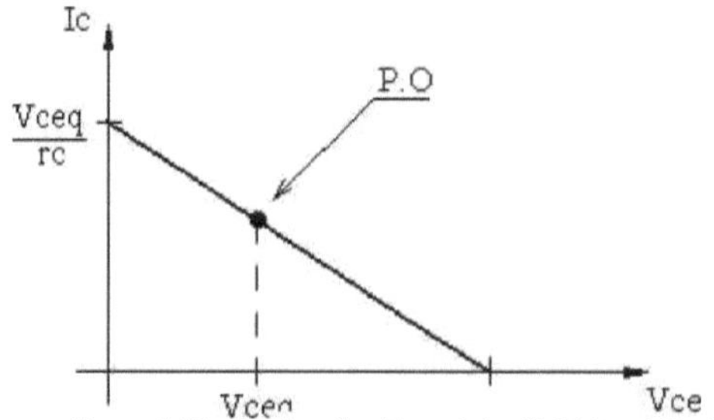

Figure 1: Bipolar Junction Transistor (BJT) Load Straight.

Operating Point (Quiescent): For transistor amplifiers, the resulting DC current and voltage establish an operating point on the characteristic curves that define the region that will be employed for amplifying the applied signal.

3 . EQUIPMENT AND PERMANENT MATERIALS REQUIRED

The following equipment and permanent materials were needed to carry out this experiment:

- Computer;
- PROTEUS software for simulating electrical and electronic circuits.

4 .EXPERIMENTAL PROCEDURE

Firstly, the circuit in Figure 2 was simulated in the Proteus software. The circuit was then fed with a voltage source equal to 12V and the function generator was set to a signal with a frequency equal to 500Hz and 5mVp.

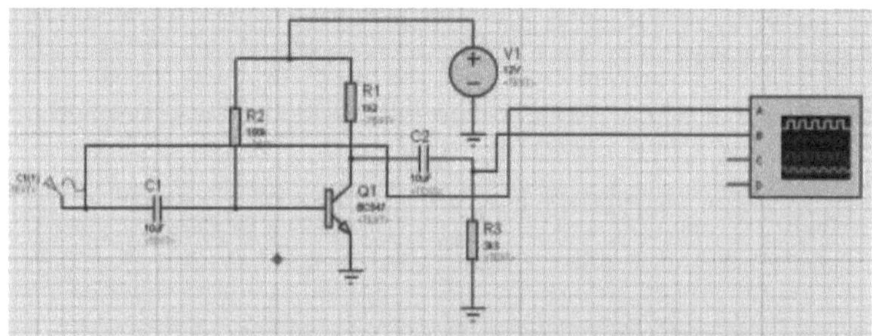

Figure 2: Circuit assembled and simulated in Proteus software.

The dc base current was then calculated using the base loop equation, remembering that the capacitor for direct current behaves like an open circuit, as follows:

$V1 - R_B \times I_B - V_{BE} = 0$ considering $R_B = R2$

$12 - 180k \times I_B - 0.7 = 0$

$I_B = 11.3/180k$

$I_B = 62.78uA$

In addition, the maximum collector current (ICsat) was calculated, considering VCE equal to zero, i.e. Ic = Vcc/Rc, as follows:

$V1 - R_C \times I_C - V_{CE} = 0$ considering $R_C = R_i$

$12 - 1.2k \times I_C - 0 = 0$

$I_C = 12 / 1.2k$

$I_C = 10mA$

The load line in Figure 3 was also drawn, adopting the following points: $(V_{CE} = 0, I_{csat})$ and $(V_{CE} = Vcc, I_C = 0)$.

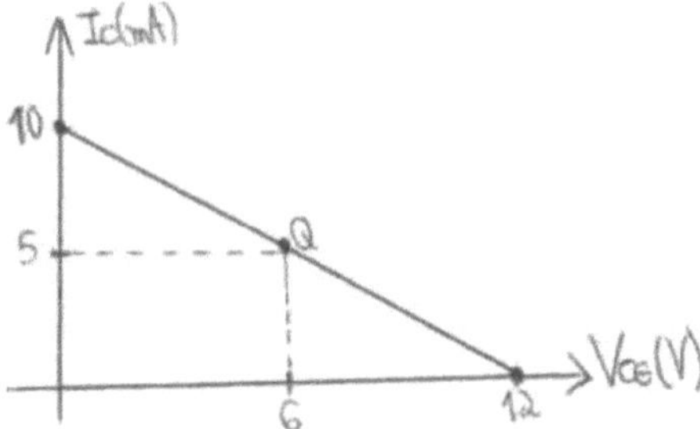

Figure 3: Load line.

Next, the function generator was set to a signal with a frequency equal to 1kHz and 5mVp. The output voltage on resistor R3 was then measured and the input and output signals are shown in Figures 4 and 5 respectively.

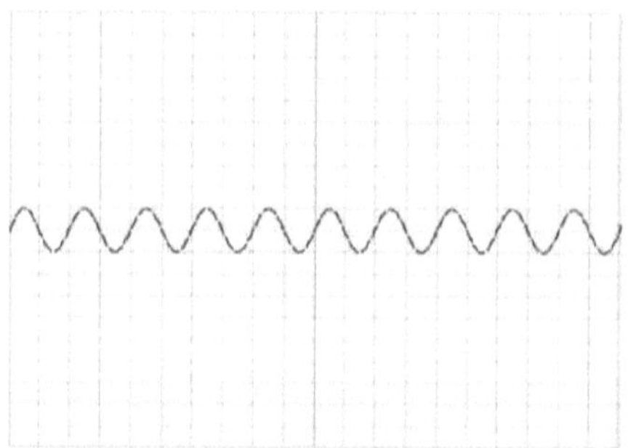

Figure 4: Oscilloscope input signal (1 A) - V/div = 5mV/div - Time/s = 500µs/div.

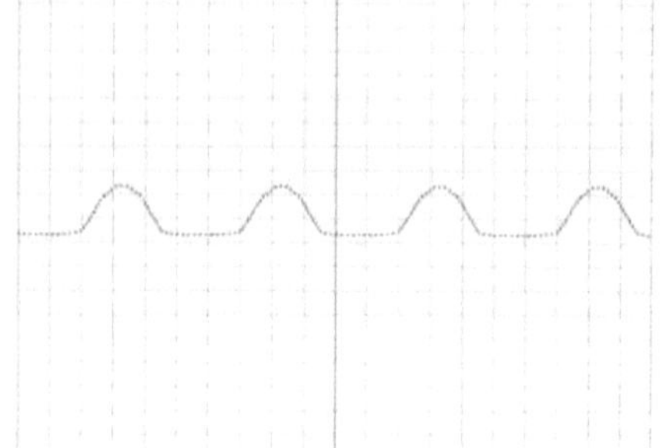

Figure 5: Oscilloscope output signal (1B) - V/div = 500mV/div - Time/s = 200µs/div.

Based on the voltage values found, the voltage gain, which represents the amplifier's performance, was calculated using the formula $G_V = V_{OUT}/V_{IN}$:

Gv = 1OOOmV /1 OmV -> Gv = 100

In addition, the value of V_{IN} was changed to 50mV and 500mV, noting that as this value increased, it got closer to becoming practically equal to the horizontal line corresponding to the time axis. As a result, the output signal in the first case increases its amplitude considerably, as well as suffering signal distortion. In the second case, the output signal is not only distorted, but also inverted, so that its Vp is turned downwards, i.e. negative. As a result, the voltage gain decreases. This is because the signal supplied to the amplifier's input produces an amplified output signal that increases with an increase in the input signal up to a certain limit. A signal with a high amplitude can produce saturation or clipping of the transistor, or both, and then the signal is clipped at the bottom or top, or both.

The V_{IN} frequency was then changed to 100kHz and it was observed that the output signal became jittery, but remained with the same initial settings. This is because the change in the amplifier's voltage gain is produced as a result of the circuit's unequal response to signals of different frequencies. This phenomenon is produced due to the presence of reactive components in the circuit, which have frequency-dependent behaviour (capacitances

and inductances). In the medium frequency range, the gain is maximum. In the low frequency range, the coupling capacitances of the circuit produce the attenuation of the output signal, while in the high frequency range, the attenuation is produced by the intrinsic capacitances of the transistor and the parallel capacitances of the circuit.

On the load line with no current through the transistor, the voltage drops across the BJT and, on short-circuit, the voltage is shared by the collector and emitter resistors.

The curves found are device-dependent, however, so you can't rely on the particular transistor to behave as described in the data sheet. That's why the load lines are really best used for developing physical understanding.

The collector output resistor connects the amplifier's output to its input, thus destroying the one-sided nature of the amplifier and complicating the analysis considerably. However, since the collector resistor is high, its inclusion in the analysis has little effect on the amplifier's performance and can be eliminated from the circuit.

Thus, by increasing the resistance of the collector resistor, its current decreases and, consequently, the angle of inclination of the load line also decreases. And when you reduce its resistance, the opposite happens. Changing the resistance of the base resistor will not interfere with the charging line, since it depends on Ic and V_{CE}, but the quiescent point will change due to the value of I_B found. However, by increasing the value of the base resistor, the signal is attenuated, i.e. the distortion of the amplifier's output signal is reduced.

The output signals at the base and collector were then checked on the oscilloscope and shown in Figures 6 and 7 respectively.

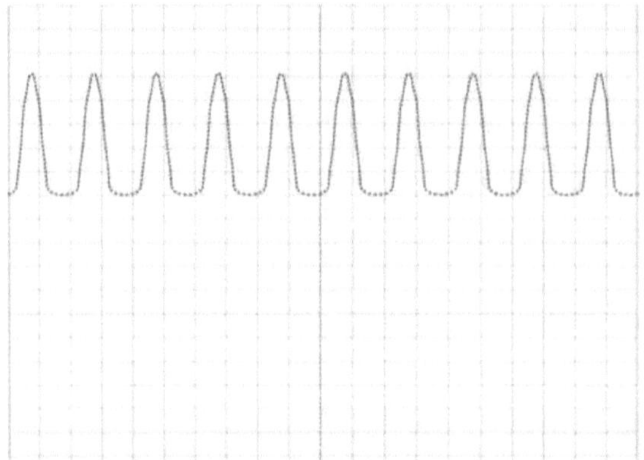

Figure 6: Output signal at the base of the transistor (1C) - V/div = 200mV/div - Time/s = 500µs/div.

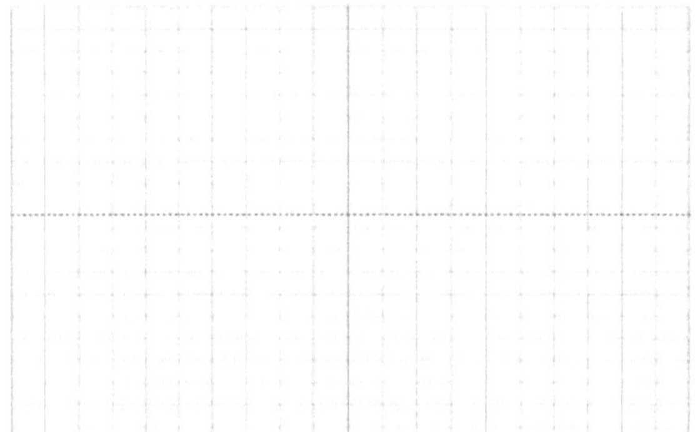

Figure 7: Output signal at the transistor collector (1D) - V/div = 20V/div - Time/s = 500µs/div.

Finally, tests were carried out by increasing the value of the base resistor in order to attenuate the output signal, i.e. so that it no longer distorts. After several attempts, this goal was achieved by changing the value of the R2 = 180kQ resistor to R2 = 90MQ, as shown in Figure 8. And Figure 9 shows the output signal found after changing this base resistor, where there are no more distortions in the amplifier's output signal.

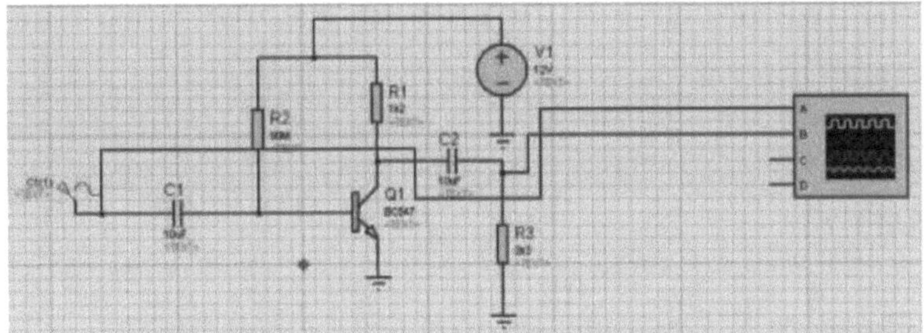

Figure 8: Circuit assembled and simulated in Proteus software with Rz = 90MΩ

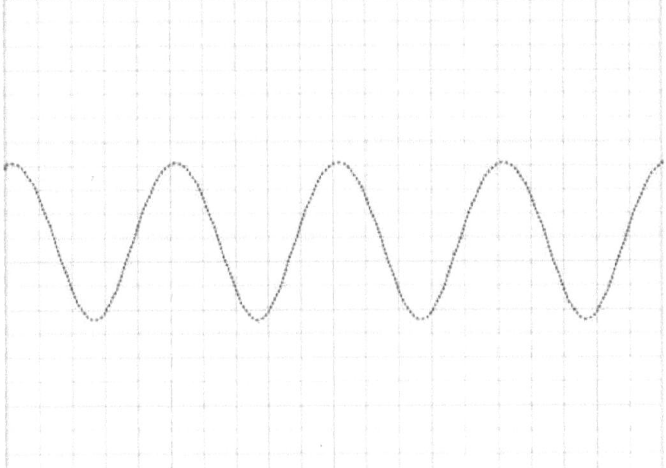

Figure 9: Output signal with Ra = 90Mĺ1 - V/div = 2mV/div - Time/s = 200µs/div.

5 . RESULTS AND CONCLUSIONS

Therefore, the objective of the experimental lesson was achieved, as we acquired theoretical and practical knowledge about the basic functioning of the bipolar transistor, as well as learning how to use it and operate it correctly when assembling the circuit in the Proteus software, which is fundamental for continuing with the syllabus of the Electronics subject.

Therefore, the assemblies and measurements requested in the script for this practical lesson were successfully carried out, and the results obtained The results of the experiment were as expected, as we learnt how the transistor works as an amplifier in an electrical circuit, acting with and without distortion. We also learnt the importance of choosing the operating point, the effect of the slope of the load line on the voltage gain and how to attenuate the output signal, i.e. eliminate the amplifier's distortion by increasing the value of the base resistor's resistance.

6 . BIBLIOGRAPHY

BOGART Jr, **Theodore F, Electronic Devices and Circuits,** Volume I, 3º edition, São Paulo: MAKRON Books, 2001

BOYLESTAD, Robert Louis; NASHELSKY, Louis. **Electronic Devices and Circuit Theory.** 6 ed. vol. II. Rio de Janeiro: Prentice Hall, 1996.

_ . **Datasheet of the BC547.** Available at: < http://www.play.com.br/datasheet/BC547.pdf >. Accessed on: 04 April 2013.

MALVINO, Albert Paul. **Electronics.** 4 ed. São Paulo: Makron Books, 1989.

SOUZA, Vitor Amadeu. **Getting to know Proteus.** Available at: <www.cerne-tec.com.br>. Accessed on: 20 April 2013.

CHAPTER 4

OPERATIONAL AMPLIFIER - INVERTER

1. OBJECTIVES

- Introduce basic concepts about operational amplifiers;
- Perform assembly using an operational amplifier as an inverter amplifier.

2 .THEORETICAL CONCEPTUALISATION

An amplifier is any device whose main function is to obtain a certain gain, i.e. the output signal is greater than the input signal [V0 = KVi].

The operational amplifier (AMPOP) is an amplifier made up of a set of transistor circuits (integrated circuit with several elements), the model of which is shown in Figure 1.

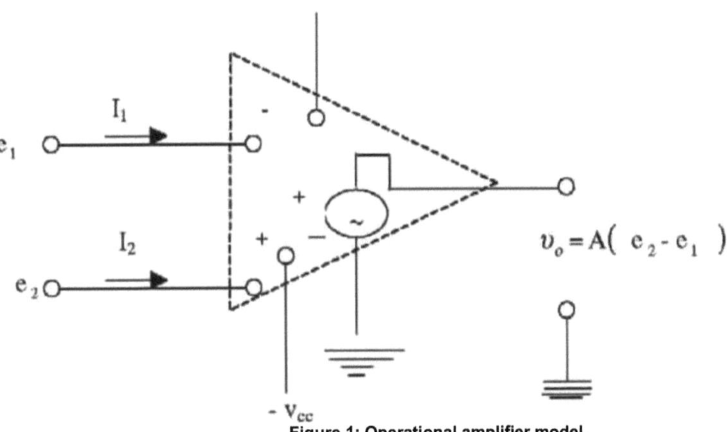

Figure 1: Operational amplifier model.

The ideal AMPOP is characterised by very high gain, very high input impedance and very low output impedance. It can be used in various configurations to generate circuits:

- Amplifiers (inverter and non-inverter);
- Adders;
- Integrators;
- Derivatives, among others.

3. EQUIPMENT AND PERMANENT MATERIALS REQUIRED

The following equipment and permanent materials were needed to carry out this experiment:

- Oscilloscope OS-21 - 2 channels;
- Tips - 2 pairs (red and black);
- Voltage source;
- Signal generator;
- 10kQ and 56kQ resistors (1 unit of each);
- Protoboard;
- CI LM741 (1 unit);
- Nose and cutting pliers, wires;
- Multimeter;
- Function generator;
- Cables for connections.

4.EXPERIMENTAL PROCEDURE

Firstly, using the available AmpOp (CI LM741), the following circuit (Figure 2) was assembled on the protoboard.

V1

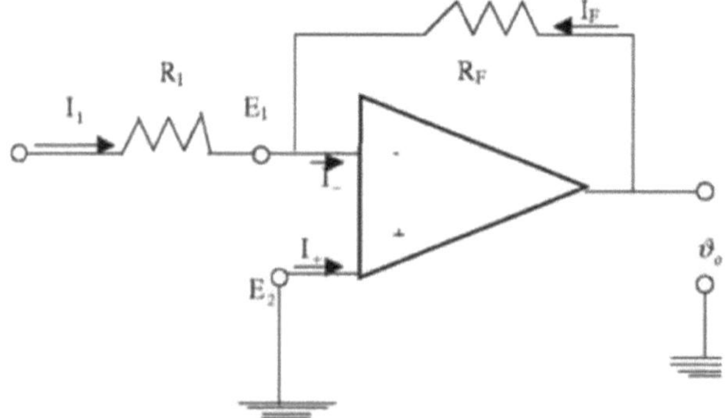

Figure 2: Circuit assembled on the protoboard.

The AmpOp available for assembly (CI LM741) must be connected

30

according to the connection diagram shown in Figure 3.

For polarisation, +VDC = 15V and -VDC = -15V were used.

RF= 50 kQ and R1= 10kQ V1=1.cos(103..ir.t)

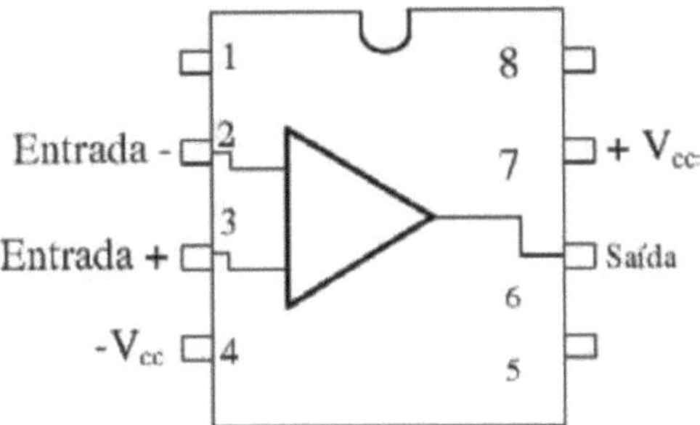

Figure 3: Connection diagram for CI LM741.

After assembling the circuit, the signals at V1 and at the Vo output were measured with an oscilloscope. The signals found are shown in Figure 4 below.

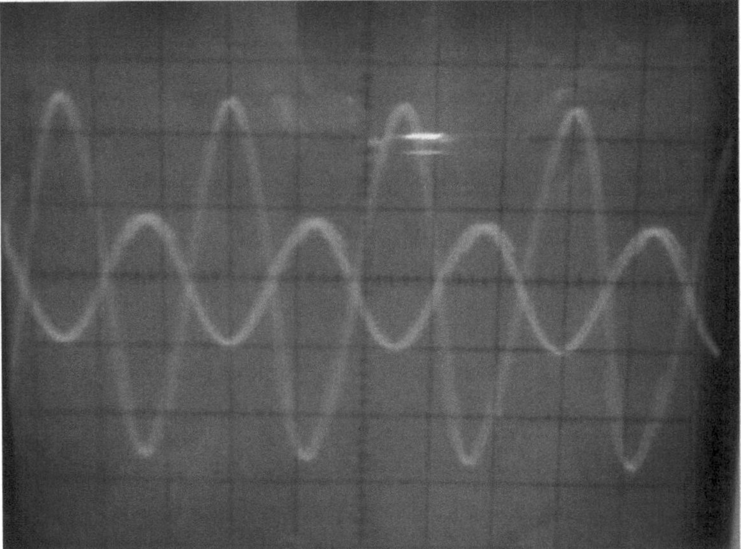

Figure 4: Oscilloscope input signal (1 A) - V/div = 1V/div - Time/s = 2ms/div.
Output signal on oscilloscope (1B) - V/div = 2V/div - Time/s = 2ms/div.

If you look at the input and output signals using the two channels of the oscilloscope, you can see that the signal has been amplified with a gain of 5V.

Next, we considered the V1 signal with 5Vp, and observed the output signal, noting that the signal had begun to saturate, i.e. distort, since the signal

31

has the maximum voltage allowed so that this distortion does not occur. If it exceeds the maximum voltage (saturation voltage) stipulated and allowed, according to the device's data sheet, the distortion seen in this situation will occur.

In addition, the CI LM741 data sheet was consulted, where its characteristics were found, such as:

- Supply voltage: ± 22V;
- Power dissipation: 500mW;
- Differential input voltage: ± 30V;
- Input voltage: ± 15V;
- Operating temperature: -55°C to 125°C;
- Storage temperature: -65°C to 150°C.

The CI LM741 data sheet also revealed the function of the pinout not used in the assembly of this experimental procedure, since pins 1, 5 and 8 have a null offset function, i.e. they can be used to cancel out the compensation seen at the output due to inadequacies in the differential circuit.

The offset function, also called the Input Offset Voltage, acts as a differential signal applied to the AmpOp's inputs and produces a differential voltage at its output (proportional to the gain of A). This output voltage is called the Output Offset Voltage (or output error voltage) and is represented by Vo (offset).

In high-precision circuits, it is necessary to minimise or eliminate this error voltage at the output of the device. In the case of an AmpOp, this error voltage is cancelled or balanced by means of a voltage divider connected to the differential input stage. This voltage divider will allow the base and collector currents to be balanced in such a way that the difference between the VBE1 and VBE2 values cancels out. This adjustment must be made with the inverter and non-inverter inputs.

connected to ground. After balancing, the desired circuit can be assembled, taking care not to alter the adjustment made.

Some AmpOps have their own terminals for adjusting the output offset voltage. However, there are others that don't have these terminals and the user will have to set up an external circuit conveniently connected to the AmpOp's inputs in order to carry out the adjustment.And finally, the gain equations for the inverter and non-inverter amplifiers were researched, which can be seen, respectively, below in Figure 6.

$$A_V = \frac{Vo}{Vi} \qquad A_V(dB) = 20\log\frac{Vo}{Vi}$$

Figure 6: Gain equations for inverting and non-inverting amplifiers.

5. RESULTS AND CONCLUSIONS

Therefore, the objective of the experimental lesson was achieved, as we acquired theoretical and practical knowledge about the basic operation of the inverter operational amplifier, as well as learning how to use and operate it correctly when assembling the circuit, which is fundamental for continuing with the syllabus of the Electronics subject.

Therefore, the assemblies and measurements requested in the script for this practical lesson were carried out successfully, and the results obtained from the experiment were as expected, as we learnt how the transistor works as an inverting operational amplifier in the electrical circuit. We also learnt about the pinout function of the LM741 integrated circuit and its main physical and operating characteristics, as well as the operation of the offset null, the input and output offset voltage, and the gain equation for inverting and non-inverting amplifiers.

Finally, the practical procedure was carried out as a team, since there are several tasks and stages to be followed. By dividing up the roles, the work is more effective and the environment is more harmonious and fruitful for developing the proposed activities and ensuring that they are carried out on time, especially when the members of the group have a good relationship with each other, and learning becomes more effective and dynamic, since all the members share knowledge and experiences relating to the subject.

6. BIBLIOGRAPHY

BOGART Jr, **Theodore F, Electronic Devices and Circuits,** Volume I, 3º edition, São Paulo: MAKRON Books, 2001

BOYLESTAD, Robert Louis; NASHELSKY, Louis. **Electronic Devices and Circuit Theory.** 6 ed. vol. II. Rio de Janeiro: Prentice Hall, 1996.

_ . **Datasheet of the LM741.** Available at: <http://www.play.com.br/datasheet/LM741.pdf>. Accessed on: 04 April 2013.

MALVINO, Albert Paul. **Electronics.** 4 ed. São Paulo: Makron Books, 1989.

CHAPTER 5

FET/MOSFET

1. WHAT?

FET stands for Field *Effect Transistor,* which, as the name suggests, works through the effect of an electric field on the junction. This type of transistor has many applications in the area of amplifiers (operating in the linear area), in switches (operating outside the linear area) or in current control over a load. The main characteristic of FETs is their high input impedance, which allows them to be used as impedance adapters and can replace transformers in certain situations. They are also used to amplify high frequencies with a higher gain than bipolar transistors.

FETs can be made up of germanium or silicon combined with small amounts of phosphorus and boron, which are "doping" substances (i.e. substances that alter the electrical characteristics). Silicon transistors are the most widely used today, while germanium transistors are only used to control high powers.

A general-purpose FET has three terminals: gate, source and drain, which allow for six forms of polarisation, three of which are the most commonly used: common source (source connected to input and output simultaneously), common gate (gate connected to input and output simultaneously) and common drain (drain connected to input and output simultaneously).

FETs can be divided into two categories: JFETS and MOSFETS, and their basic characteristics are the control of a current by an applied electric field. The current flows between the terminals, called Supply - S, and Drain - D, and the field due to a voltage applied between a control terminal, the "Gate" - G, and the supply. This compartment is analogous to that of pentode electronic valves.

The practical advantage of FETs, which is making them more and more common, especially MOSFETs, is their high input inpedance: practically no input current is needed at the gate to control the drain current.

The MOSFET transistor (acronym The TECMOS (Metal Oxide Semiconductor Field Effect Transistor) is by far the most common type of field effect transistor in both digital and analogue circuits. Its basic principle was first proposed by Julius Edgar Lilienfeld in 1925.

The word "metal" in the name is an anachronism from the first chips, where the gates were made of metal. Modern chips use polysilicon gates, but are still called MOSFETs. A MOSFET is made up of a channel of N-type or P-type semiconductor material and is called NMOSFET or PMOSFET respectively. Generally, the semiconductor of choice is silicon, but some manufacturers, notably IBM, have begun to use a mixture of silicon and germanium (SiGe) in the channels of MOSFETs. Unfortunately, many semiconductors with better electrical properties than silicon, such as gallium arsenide, do not form good oxides in the gates and are therefore not suitable for MOSFETs. IGFET is a related term that stands for Insulated-Gate Field Effect Transistor, and is almost synonymous with MOSFET, although it can refer to a FET with a gate insulated by a non-oxide insulator.

The gate terminal is a layer of polysilicon (polycrystalline silicon) placed over the channel, but separated from the channel by a thin layer of insulating silicon dioxide. When a voltage is applied between the gate and source terminals, the electric field generated penetrates through the oxide and creates a kind of "inverted channel" in the original channel below it. The inverted channel is of the same P-type or N-type as the source or drain, so it creates a conductor through which the electric current can pass. Varying the voltage between the gate and the source modulates the conductivity of this layer and makes it possible to control the current flow between the drain and the source.

2. **CONSTRUCTION**

The first FET developed was the junction FET (Junction Field Efect Transistor). There are two types: N-channel and P-channel, as can be seen in

Figure 1.

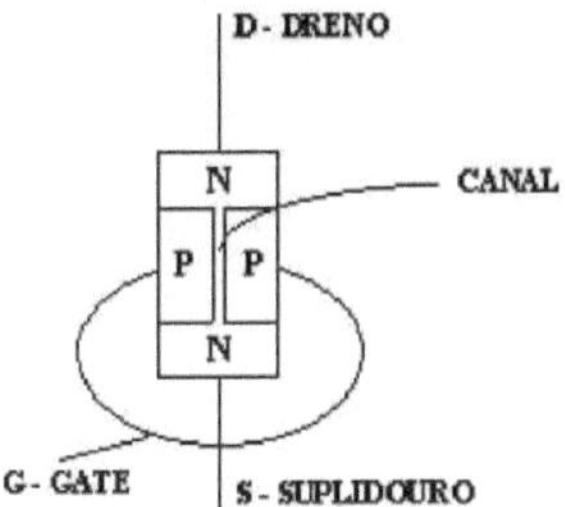

Figure 1: MOSFET structure - N channel.

Its structure consists of a bar of N (or P) semiconductor material, surrounded in the centre by P (or N) material. This narrow part of the N (or P) region is called a channel because it influences the controlled current. In the P-channel FET, the N and P semiconductor layers are reversed. The schematic symbols of the N and P channel FET are shown in Figure 2.

Figure 2: FET schematic symbols.

Note that a potential region forms around one channel at the PN junction. This barrier restricts the conduction area from one channel to the other.

3. OPERATION

In Figure 3 below, we have the MOSFET test circuit with a variable source Ves, which controls the ID channel current. Note that Ves is in reverse polarisation (- at gate P).

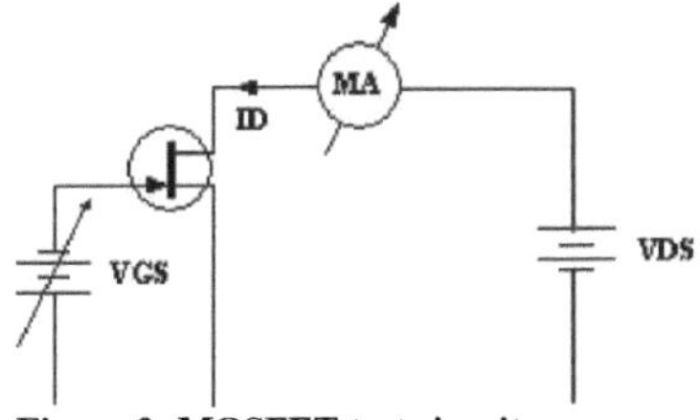

Figure 3: MOSFET test circuit.

Initially we make Ves = 0. The N channel is normally open, as the power barrier is minimal, so a maximum current, called IDSS, is flowing, characteristic of the MOSFET for Vds.

Now let's increase Ves, causing the width of the potential barrier to increase. Then the conduction area decreases, which also decreases the drain current. The electric field between the gate and the supply repels electrons from the channel near the junction and the current is confined to the centre, decreasing. This is the field effect, which gives the transistor its name.

The higher the reverse voltage Ves, the lower the drain current, with a fixed Vds. If we gradually increase it, there will come a point where the current cancels out. The voltage Vgs at this point is called Vgsoff or Vgscorte, the channel choke or cut-off voltage.

4. MAIN SETTINGS

There are two types of characteristic curves for the MOSFET: transconductance and drain, shown in Figures 4 and 5 respectively.

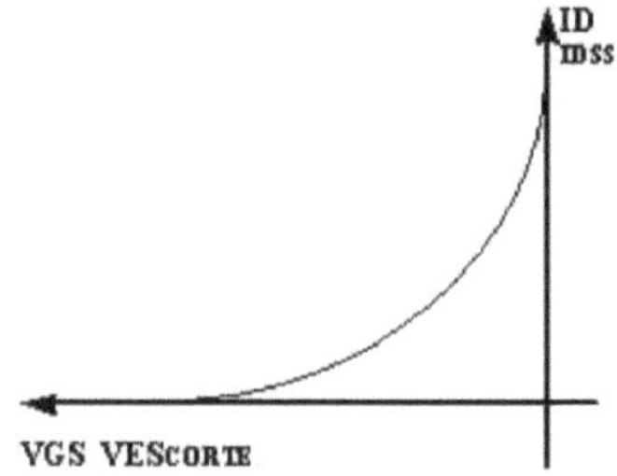

Figure 4: Transcundance characteristic curve.

This curve, valid for Vds > Vgs cut-off, describes the control of the drain current by the gate/sink voltage. It is the MOSFET's active region curve.

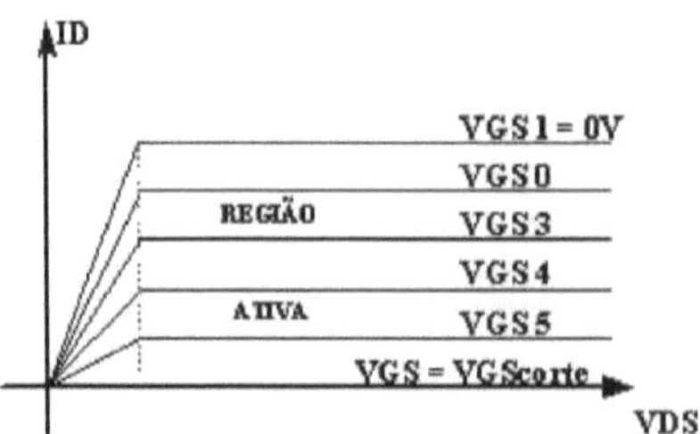

It is analogous to the collector characteristic of a bipolar transistor, and similar to the plate characteristic of a pentode valve. It describes the behaviour in the three operating regions, for different values of Vgs.

In the active region, the drain current is controlled by the voltage Vgs, and hardly varies with the voltage Vds (controlled current source compartment). In this region, the MOSFET can function as a current-source multiplier.

The MOSFET is in this region when Vds > Vescut in the characteristic curves is the horizontal part of the curve for a certain Vgs (all the area outside saturation, hatched, and between the curves Vgs1 and Vgs6).

Saturation occurs when Vds < Vgscorte. Here the ID current depends on both Vgs and Vds (controlled resistor behaviour). In the drain characteristic curves, it is the sloping line that joins each curve to the origin of the graph. Note that the slope, related to the channel resistance, is different for each of the curves (Vgs values). In this region, the MOSFET acts as a voltage-controlled resistor or switch, depending on the application.

When Vgs || Vgscut, the MOSFET is in the cut-off region, and the drain current is zero. Used in switch operation (alternating with saturation - closed switch).

5. WHICH ONE TO USE: FET/MOSFET/TRANSISTOR?

The FET has several advantages over the bipolar transistor:

- Their operation depends only on the flow of majority carriers. They are therefore unipolar devices (they only work with one type of charge carrier) and are therefore also called unipolar transistors;

-They are   relatively immune to radiation;

-They have   a high input   resistance, typically of the order of MQ;

-They are   less noisy than bipolar transistors;

-They   have no residual voltage (offset voltage) for zero drain current;

-They are   thermally stable.

The disadvantage of the FET is its small gain x bandwidth product compared to the bipolar transistor.

In practice, MOSFET technology is used in electronic equipment to amplify the signal sent by the main unit to power the speakers. This results in greater power for the sound produced. And that's not all. In addition to greater power, MOSFET technology influences the quality of the audio produced, avoiding distortion and contributing to better sound efficiency in the device.

There are also operational amplifier models based on FET/MOSFET technology, which are very useful and widely used in the electronics industry.

6. APPLICATIONS

A) Current Source (Figure 6):

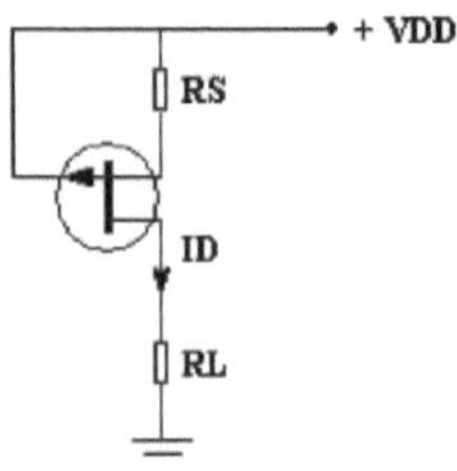

Figure 6: MOSFET as a current source in a circuit.

The value of RS and the curve of the MOSFET determine the ID current.

The circuit operates the MOSFET in the active region, i.e. Vds> Vgscorte, which imposes a limit on the value of RL.

The circuit is used in polarisation, often within operational amplifiers and other analogue CI's.

B) Amplifiers:

When operating as amplifiers, we use the concept of transconductance, which defines the gain of FETs.

$$gm = \gamma = \frac{\Delta ID}{\Delta VGS}$$

Transconductance, gm, is the ratio between the variation in current Id and the variation in Vgs that causes it.

In FETs, the transconductance is higher for lower bias voltage Vgs and higher current ID.

Thus, the gain is determined by the polarisation, as in bipolars and valves, and the type of FET.

7. SIMULATION

In order to improve the knowledge acquired about FET/MOSFETs, a simulation was carried out in the Proteus software (Figure 7), using a MOSFET of the IRC634 type as a DC switch, which is much easier, more practical and efficient than using a BJT with the best characteristics. To do this, the datasheet of the MOSFET used was researched.

The switch works as follows: activating the input clock on the MOSFET switches the lamp on and, consequently, deactivating it switches the lamp off. The voltage and frequency values were tested until satisfactory results were obtained.

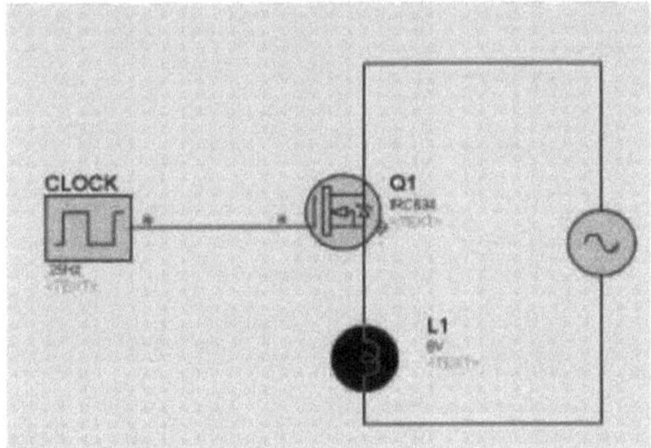

Figure 7: Simulation of a DC switch using MOSFET.

REFERENCES

BOYLESTAD, Robert Louis; NASHELSKY, Louis. **Electronic Devices and Circuit Theory.** 6 ed. vol. II. Rio de Janeiro: Prentice Hall, 1996.

MALVINO, Albert Paul. **Electronics.** 4 ed. São Paulo: Makron Books, 1989.

Printed by Books on Demand GmbH, Norderstedt / Germany